The Loco
of Sir Nigel Gresley
1911 – 1921

By

O. S. Nock

British Library Cataloguing-in-Publication Data
A catalogue record for this book is available from
the British Library

Contents

*Introduction—" J6 " 0-6-0s—" J2 " 0-6-0s—" K1 " and " K2 "
Moguls—" O1 " 0-8-0s*

THE career of Sir Nigel Gresley has no parallel in British locomotive history. First as Locomotive Superintendent of the Great Northern Railway, and then as Chief Mechanical Engineer of the L.N.E.R., his chieftainship bridges not merely 30 years of great events in the railway world, but has also witnessed a complete transition from the old operating methods, when engines were nursed and groomed like racehorses, to the stringent economic conditions of today. It links the time when single-wheelers were still being used on crack expresses to our modern streamline age, when up to the outbreak of war developments both at home and overseas were following thick and fast upon each other.

Throughout G.N.R. and L.N.E.R. history there has been a marked continuity in the locomotive design of these companies. The streamlined Pacifics are lineal descendants, not merely of the large-boilered Ivatt Atlantics, but of far older types—the Stirling 8-footers, and, even before them, the celebrated Sturrock 4-2-2, No. 215. The big changes that often take place on the appointment of a new chief are absent in Great Northern history, yet each engineer, in developing the practice of his predecessor, has not been content merely to keep abreast of traffic requirements. In the early years of each *régime* an outstanding express design has been produced which with various refinements has remained the standard type for very many years.

Stirling's was the 4-2-2 era ; during Ivatt's reign, although singles and 4-4-0s were regularly employed in express service, one could not regard the Great Northern as anything but a 4-4-2 line, and Sir Nigel Gresley's day is that of the Pacific. The arrival on the scene of the 4-6-2s was delayed by the last war, but the experience gained during the war in the haulage of enormous loads—by pre-1914 standards—was probably of no small value in the preparation of the Pacific design. Again although one wheel arrangement has remained standard through-

1

[Loco. Publishing Co.

Photo]

At the start of the Gresley régime. The 2.15 p.m. Leeds and Bradford Flyer headed by 2-2-2 locomotive No. 872 G.N.R.

out each era, the pioneer design in each case has been materially improved during the career of its creator. One would no more think of equating the merits of a non-superheater " Klondyke " to those of a " 251 " of today as one would place the *Great Northern* of 1922 in the same power classification as *Golden Eagle* or *Silver Link*.

In 1911 the Great Northern was a more interesting line to a lover of historic locomotives than to a train-running enthusiast. If the lightly-loaded crack expresses, such as the 2 p.m. from Leeds to Kings Cross, and the short-lived 2.15 p.m. down Yorkshire flyer, be omitted, the majority of the main line services, and the Scotch expresses in particular, had a most leisurely collection of timings. The Ivatt Atlantics, then without superheaters, were indifferent in their performance, and although never piloted, often fell to but little over 30 m.p.h. with 300-350 ton loads up the long 1 in 200 banks, with the result that lost time was common. More entertaining running occurred on trains hauled by 4-4-0s and the various single-driver types ; 4-4-0s were often used on the Scotch expresses north of Grantham, and in 1910-11 the 5.30 p.m. down Newcastle diner, usually only six coaches beyond Grantham, was almost invariably hauled by a single over this section. The four-coach 2.15 p.m. down, which covered the 156 miles from Kings Cross to Doncaster in 165 min., was usually worked by a Stirling 2-2-2, though the famous 2 p.m. up from Leeds, normally five or six cars, was hauled by an Atlantic ; this train had both the longest and one of the fastest runs on the G.N.R.—Wakefield to Kings Cross, 179.8 miles, in 187 min. The fastest train was the up Manchester express leaving Grantham at 7.48 p.m., which was allowed 110 min. for the 105.5 miles to Kings Cross ; but the load rarely exceeded four coaches.

Owing to the agreement between the East and West Coast companies the day Scotch expresses were timed very slowly over the Great Northern ; the 10 o'clock or Flying Scotsman took 120 min. from Kings Cross to Grantham, and 98 min. for the 82.7 miles thence to York. The 2.20 p.m., pre-war counterpart of the 1.20 p.m. of 1932-9 days, took 122 and 97 minutes respectively over the two stretches. The ordinary Leeds trains did some considerably smarter running, including a Kings Cross-Peterborough run in 82 minutes—55.9 m.p.h. average—but their loads were usually not more than 250-300 tons. The Leeds

The 5.54 p.m. stopping train from Kings Cross near Hadley Wood. Class "536" 0-6-0 locomotive No. 3602 L.N.E.R. (G.N.R. No. 602)

[E. R. Wethersett

Photo]

and Manchester flyers were in a class apart, much as the stream-lined trains of later years ; ordinarily they kept good time.

But in October, 1911, the date of Gresley's appointment as Locomotive Superintendent of the G.N.R., it was not in express passenger motive power that the greatest need lay. A marked trend towards faster goods services all over the country was evident ; the G.N.R. was in the forefront of this movement, but strangely enough the company did not possess any engine really suitable for the traffic. In consequence every available passenger tender engine was pressed into service, and Atlantics, 4-4-0s, Stirling 2-4-0s, and even singles could be observed working freight trains. It was this boom in fast goods traffic that began to break down the old principle of " one driver, one engine," for the fullest use had to be made of locomotives such as the large Atlantics, most of which included a fair proportion of fast goods mileage in their regular rosters. But the use of other types could be regarded only as a temporary measure pending the construction of suitable engines. Such was the Great Northern's need that for the first ten years of Mr. Gresley's chieftainship at Doncaster all new engines turned out were intended for freight service.

The first new type to appear was actually an Ivatt design, a 0-6-0 superheater goods, with 5 ft. 2 in. wheels, and having the same fine-steaming boiler as the " 51 " class 4-4-0 express engine. They had cylinders 19 in. × 26 in., and although the boiler was dimensionally the same, having 1,230 sq. ft. of heating surface, the working pressure was 170 lb. per sq. in. against 160 lb. in the 4-4-0s. The first batch, of 15, came out in 1911 and were numbered 521-535 ; by successive additions, which continued up to 1921, the class now musters 110 strong. Their present numbers are 3521-3610, and 3621-3640, and they are now L.N.E.R. class " J6." The later examples have Robinson superheaters.

The " 536 " class, as the foregoing engines were usually called, were followed early in 1912 by another goods locomotive design. This, now class " J2," was also a superheater 0-6-0, but having wheels no less than 5 ft. 8 in. in diameter. In this respect they were similar to Ivatt's No. 1 class, built in 1908, but the boiler, cylinders and motion were the same as that of the " 536 " class. Ten were turned out, Nos. 71-80, and were immedi-ately successful. They worked the York-London through goods nightly, and also had a long turn from Peterborough to Manchester. Like the " No. 1 " class they were real mixed traffic engines, and at times of pressure took turns on passenger working, excursions and such like, which in those days were not timed at particularly high speeds.

But a 0-6-0 locomotive is not ideal for duties needing speeds of 60 m.p.h. or more, no matter how well aligned the road may be. In preparing a new design to meet contemporary needs, therefore, Gresley followed the general trend of British practice at the time for mixed traffic work, and chose the Mogul wheel arrangement. Since the advent in 1899-1901 of the imported 2-6-0s on the Midland, Great Central, and Great Northern, the Mogul was in many quarters looked upon as an undesirable Yankee intrusion. By the year 1912, however, its popularity had been firmly assured by Churchward's " 43XX " class on the Great Western, and the new Great Northern engines appeared almost contemporaneously with 2-6-0s on the Brighton, Caledonian, and Glasgow & South Western ; not many years later yet another example appeared on the S.E. & C.R. Gresley's engine was described by *The Railway Magazine* of the day as " a No. 1 class 0-6-0 with the addition of a pony truck " ; but actually the new type was a far greater departure, and as the parents of a large and successful family of engines the Moguls of 1912 are worthy of special attention.

Ten of the type were built, and numbered 1630 to 1639. Their leading dimensions were : cylinders, 20 in. × 26 in., coupled wheels 5 ft. 8 in. dia., total heating surface 1,420 sq. ft., grate area 24.5 sq. ft., working pressure 170 lb. per sq. in. The principal feature was the front end ; Walschaerts valve gear was used, working 10 in. dia. piston-valves. The valve setting was carefully arranged so as to give a large exhaust opening when the engine was running well linked-up. Apart from the outside Walschaerts gear and the high raised running plate their appearance was thoroughly Great Northern, and the footplate arrangements unaltered from standard practice since Stirling days, the characteristic feature of which was the pull-out type of regulator working in a horizontal plane.

The Moguls of 1912 were not unduly long in showing what they could do ; they worked the fast night goods to Doncaster, a " lodging " turn, and a great variety of mixed traffic jobs, including express passenger trains at times of pressure. In the

One of Gresley's first 2-6-0 locomotives, 2-cylinder No. 1636, G.N.R.; L.N.E.R. Class "K1," now 4636 Class "K2" (re-boilered)

Photo]   [W. J. Reynolds

G.N.R. No. 420 reboilered by Gresley; renumbered 3420, L.N.E.R. Class "Q3"

The boiler is similar to that fitted to "K2" Moguls

[W. J. Reynolds

Photo]

7

Photo]   [W. J. Reynolds
Gresley's larger-boilered 2-6-0, G.N.R. No. 1655, painted green
Later classed " K2" L.N.E.R., and renumbered 4655

early days of the last war they were often requisitioned for
ambulance train workings. If there was a weakness in their
design it lay in the boiler, which in proportion to the cylinder
dimensions was by Great Northern standards on the small side;
the next batch, which came out in 1914, had boilers of 5 ft. 6 in.
dia., instead of 4 ft. 8 in., though the cylinders remained the
same. No. 1640 was the pioneer of a series numbering 75 engines;
their present designation is class " K2," but on the G.N.R. they
formed class " E1." As wartime train-loads increased and
schedules were eased out, they were sometimes to be seen on
regular express passenger turns, but usually there was far too
much in the way of fast goods, munition, and troop trains for
them to be spared as deputies for the Atlantics. Indeed, they
have never seemed quite at home on express working, but as fast
freight engines they were second to none at the time of
construction.

Since the grouping took place all the " 1630s " have been
rebuilt as " K2s," and the sphere of activity of the latter class
has been greatly extended, to East Anglia and over the arduous
gradients of the West Highland line.

Great locomotive power was needed for other classes of
freight service too. The coal traffic between Peterborough and

Class " K2 " Mogul No. 1669, with original G.N.R. boiler mountings. Engine has outside steam pipes. (The earlier engines, Nos. 1640-1659, had inside steam pipes) Renumbered 4669, L.N.E.R., Class " K2 "

London had already grown to large dimensions, so much so that the Ivatt " 401 " class 0-8-0s, though satisfactory engines in themselves, were being worked at near their full capacity. Gresley's new design of 1913, the " 456 " class of 2-8-0s, was a logical development both of the previous 0-8-0s and the " 1630 "

The pioneer G.N.R. 2-8-0, No. 456, 2-cylinder type, now L.N.E.R. No. 3456, Class " O1 "

class Moguls, designed for the heaviest freight service. The pony truck, in addition to supporting a heavy and powerful front end, provided a degree of flexibility in the vehicle that is especially valuable for freight working on a route like the G.N.R. main line, with its frequent diversions from fast to slow road and vice versa.

As motive power units the " 456 " class (L.N.E.R. Class " O1 ") proved as good as their ample dimensions suggest. Their cylinders were 21 in. × 28 in., and in combination with 4 ft. 8 in. coupled wheels, and a working pressure of 170 lb. per sq. in. give the class a nominal tractive effort of 31,000 lb. at 85 per cent. of the working pressure. The boiler was the largest Doncaster had produced up to that time, the barrel being 15 ft. 5 in. long, and 5 ft. 6 in. dia., and the heating surface of the tubes alone was 1,922 sq. ft. Unlike the first Moguls, these 2-8-0s were fitted with Robinson superheaters ; a high degree of superheat was evidently aimed at, for the heating surface provided, 570 sq. ft., was unusually large for that period. The grate area of these engines was 27 sq. ft. ;, adhesion weight $67\frac{1}{2}$ tons ; and total weight of engine and tender in working order $119\frac{1}{4}$ tons. Equipment included 10-in. dia. piston valves, as in the Moguls, and the Weir feedwater heater and feed pump. In outward appearance the engines remained faithful to Great Northern traditions, and the Stirling type of regulator handle still featured among the footplate fittings. Thus in the short space of three years Gresley had produced four eminently good designs, three of which, the " 71 " class 0-6-0s, the " 1640 " class Moguls and the 2-8-0s were suitable for extensive building if the need arose. Actually they were used merely to meet the immediate requirements of the moment, while still more successful designs were being worked out.

Photo]   [W. J. Reynolds

Rebuilding of Atlantics 279, 1300 and 1421—development of 3-cylinder propulsion engine 461 and the " K3 " Moguls—" J50 " 0-6-0T — " N2 " 0-6-2 — locomotives for war service — " O2 " 3-cylinder 2-8-0s—modification of the Ivatt large Atlantics

BY the beginning of 1915 immediate needs had been met, and Great Northern locomotive history was just entering an interesting transition stage. All over the country new locomotive types were being produced; on all hands superheating was being hailed as the final and conclusive answer to the exponents of compounding, yet on the G.N.R. nothing very much seemed to be happening, outwardly at any rate, in the realm of express passenger motive power. Elsewhere one of the most strongly marked trends of the time was the introduction of multi-cylindered single-expansion locomotives. By 1914 the Great Western had practically standardised the 4-cylinder system for crack express engines ; the L.N.W.R. " Claughtons " were out and doing good work, by the standards then prevailing, while the Great Northern's own historic partner, the North Eastern, had already turned out a considerable variety of 3-cylinder simple designs.

Throughout the Gresley *régime* close study of contemporary practice, both at home and overseas, has been applied to design at Doncaster works, and in the early years of his chieftainship it is not surprising, in view of what was taking place elsewhere in the country, that Gresley made some experiments with multi-cylindered locomotives. The first step was the complete rebuilding, in 1915, of Atlantic No. 279, as a four-cylinder simple ; this engine was one of the standard Ivatt " 251 " class, having two cylinders 18¾ in. dia. by 24 in. stroke. As rebuilt the engine was provided with four cylinders 15 in. dia. by 26 in. stroke—a 40 per cent. increase in cylinder volume—and the boiler was modified by the fitting of a 24-element Robinson superheater (instead of the previous 18-element one) affording 427 sq. ft. of heating surface. This heating surface was the same as that of the original Ivatt superheater Atlantics of the 1452-1461 series. No. 279 was fitted with the Walschaerts valve-gear, but only two sets were provided, the valves of the inside cylinders being actuated by rocking shafts driven off the tail rods of the outside cylinder

Photo]   [W. J. Reynolds
No. 279 Ivatt " Atlantic," rebuilt by Gresley as a 4-cyl. simple

valve spindles. This rebuilding increased the weight of No. 279 from the original 65½ tons to 73½ tons. As rebuilt No. 279 was in nominal tractive effort the most powerful express engine on the G.N.R., though curiously enough she never came into the limelight to the same extent as the ordinary " 251 " class. In her second rebuild, in 1938, as a 2-cylinder engine with Walschaerts valve-gear, she still remains unique among the Atlantics. As now running, No. 3279 (G.N.R. No. 279) has two cylinders 20 in. ×

Photo]   [W. J. Reynolds
L.N.E.R. 3279, again rebuilt as a 2-cylinder simple, with " K2 " cylinders and motion

20

Four-cylinder compound "Atlantic" No. 1300, originally built by the Vulcan Foundry, Ltd., . .

. . . and as rebuilt as a 2-cylinder simple by Gresley in 1918

21

13

26 in., but she is not the first G.N.R. Atlantic to be so equipped. In 1917 the 4-cylinder compound No. 1300, built by the Vulcan Foundry Co. Ltd., was converted into a 2-cylinder simple by Gresley, and a pair of standard 20 in. × 26 in. cylinders, as used on the " K1 " and " K2 " 2-6-0s, was fitted. The original boiler, after twelve years of continuous service in express traffic, was retained and modified for superheated steam ; a 22-element superheater was fitted, having a heating surface of 280 sq. ft. The working pressure was reduced from the 200 lb. per sq. in. of the original compound to 170 lb. per sq. in. No. 1300, as converted, was nominally more powerful than the standard superheated Atlantics of the " 251 " class, and for a time during the early months of 1918 she was working on the 5.30 p.m. Newcastle express from Kings Cross. With a nightly load of over 400 tons, and an allowance of 125 min. to cover the 105.5 miles from Kings Cross to Grantham it was a difficult turn. Another Gresley conversion was that of the Ivatt 4-cylinder compound Atlantic No. 1421, into a standard superheater " 251 " in 1920.

Photo]   [W. J. Reynolds
Ivatt 4-cylinder compound " Atlantic " No. 1421, rebuilt by Gresley in 1920 as a standard "Atlantic" (Class " C1")

It was not until May, 1918, three years after Gresley's interesting experimental rebuilding of No. 279, that the next multi-cylindered engine appeared on the G.N.R. ; this was the first 3-cylinder 2-8-0, No. 461. The suitability of 3-cylinder propulsion

22

The first Gresley 3-cylinder locomotive, 2-8-0 No. 461

for heavy freight working had already been demonstrated on the North Eastern Railway, where Wilson Worsdell's Class " X " 4-8-0 tanks were operating successfully in the Erimus hump yard. The system equally has advantages in such duties as the haulage of the Peterborough-London coal trains, in which a locomotive may have to start heavy loads from rest against a 1 in 200 gradient. A more even crank effort is obtained with a 3-cylinder engine, having its cranks set at 120 degrees to each other, than with the 2-cylinder arrangement, in which the cranks are at right-angles to each other, and a smoother start is possible ; it is not so much a matter of power as the way in which that power is applied to the drawbar.

No. 461 carried a boiler identical with that of the " 456 " class, but the two 21 in. × 28 in. cylinders were replaced by three cylinders, 18 in. × 26 in., arranged in line and driving the second pair of coupled wheels. The connecting rods are thus much shorter than in the earlier engines, and the cylinders are steeply inclined. But at the time of construction the outstanding feature of No. 461 was the valve-gear. All previous 3-cylinder simple lcomotives built in this country—Robinson's Great Central 0-8-4 humping tanks, his one 3-cylinder simple Atlantic, No. 1090, and the various North Eastern types—used three sets of valve-gear. By the mechanism illustrated in principle in Fig. 1, however, Gresley eliminated one set of valve-motion. As applied to No. 461 the details were rather different, and are shown in Fig. 2. The cross-sectional view of the cylinders and valves shows why it was necessary to place the valve casing for the inside cylinder in a different transverse plane from that of the two outside cylinders ; this disposition involved the use of vertical levers in the derived valve-motion, and made the layout of the

23

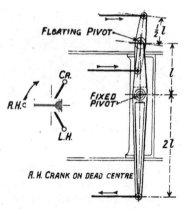

Fig. 1. *Standard Gresley derived valve-motion*

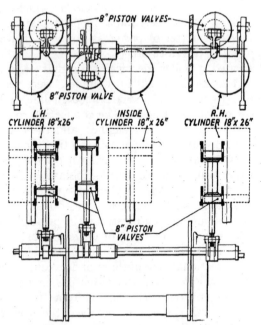

Fig. 2. *Layout of the valve-gear on engine No. 461*

24

gear rather more complicated. Later, however, the arrangement shown in Fig. 1 became the standard form of the gear. Another novel feature of this engine has since become standard on all the largest L.N.E.R. locomotives ; this is the vertical-screw reversing gear. The adjustment of cut-off is facilitated by the inclusion of ball bearings in the screw mounting.

No. 461 was put to work on the Peterborough-London coal trains and very soon showed a certain superiority over the " 456 " class, particularly in starting. In the haulage of a heavy coal train, weighing 1,300 tons gross behind the tender, a cut-off of 45 per cent., with the regulator something below one-half open, was needed to maintain an average of 22½ m.p.h. over the 15 miles from Huntingdon to Sandy, on a typical journey. In getting away, and on the 1 in 200 banks, as much as 60 per cent. cut-off was customary. But in spite of the successful working of No. 461, building of the " 456 " class 2-cylinder 2-8-0s continued.

The next 3-cylinder engine was another advance upon former practice, and its advent was surrounded by just enough secrecy to set going a flood of rumours. Soon after the armistice of November, 1918, the news got about that a " super " main line locomotive was under construction at Doncaster ; everything pointed to a Prairie at least, if not a Pacific, and then, early in 1920, No. 1000 came out, the first of the 3-cylinder Moguls. This remarkable engine created quite a stir at the time by reason

Photo]   *[W. J. Reynolds*
No. 1000, the original 6 ft. boiler 2-6-0 and the first engine to have the standard 3-cylinder derived gear. Now 4000, L.N.E.R. Class " K3 "

25

of her boiler ; up till then a diameter of 5 ft. 6 in. had been regarded as the maximum conveniently possible within the British loading gauge, whereas the boiler of the Gresley 3-cylinder Mogul was 6 ft. dia. over the smallest ring, and accommodated 217 tubes of 1¾ in. outside dia. The other leading boiler dimensions are : heating surface, tubes, 1,719 sq. ft. ; firebox, 182 sq. ft. ; 32-element superheater, 407 sq. ft. The grate area is 28 sq. ft. and the working pressure is 180 lb. per sq. in.—the latter a slight advance upon previous Gresley practice.

There were further features of interest in the front end, for not only was the cylinder volume far in excess of that of any 8-wheel engine then at work on any other British railway, but there was also an alteration in the arrangement of the valve-gear. The three cylinders were 18½ in. dia. × 26 in. stroke, and by inclining the central cylinder at a much steeper angle than the outside ones it was possible so to arrange the valves that the simple horizontal rocking lever mechanism (Fig. 1), subsequently standardised, could be used for actuating the piston-valve of the inside cylinder. This was found to be a great improvement upon the layout of the gear used on the 2-8-0 engine No. 461. Another detail destined to become standard practice on the L.N.E.R. which made its first appearance on No. 1000 was the provision of twin regulator handles, one on each side of the cab ; the handle on the fireman's side is often of great value when a locomotive is being manoeuvred in a busy yard. The outward appearance of the cab remained faithful to Great Northern traditions, though nowadays of course only a tiny minority of ten in the great regiment of " K3 " Moguls possess this outward sign of true Doncaster lineage. The original engines, though intended mainly for fast goods working, came into the limelight during the coal strike of 1921, when they tackled express passenger trains loading up to 20 bogie vehicles on the fastest schedules then operating between Kings Cross and Doncaster, and showed themselves capable of 75 m.p.h. on stretches like that from Stoke summit down to Peterborough.

In the meantime much important progress was being made in the modernisation of existing types. Quite apart from the super-heating of the " 251 " class Atlantics, to which special reference is made later, the performance of a number of Ivatt types was improved. Larger boilers, with superheaters, were fitted to some of the 4-4-0 express locomotives ; a number of the small Atlantics

26

On the G.N.R. near New Southgate in 1921. (The 5.40 p.m. express ex Kings Cross hauled by a 3 - cylinder 2-6-0 No. 1003)

Photo] [H. Gordon Tidey

No. 1003, 3-cylinder 2-6-0, now L.N.E.R. Class "K3" No. 4003

27

19

G.N.R. No. 157, Gresley's 0-6-0 shunting tank engine of 1913, the forerunner of L.N.E.R. Class "J50." Renumbered 3157, L.N.E.R. Class "J51," now Class "J50"

L.N.E.R. No. 3221, the standard "J50" with large bunker

28

Photo]

[W. J. Reynolds

and the 0-8-0 goods engines were also superheated, and one of the most interesting conversions was that applied to the Ivatt 0-8-2 tanks. After their departure from the London suburban district these engines were transferred to Colwick for use in the coal and goods marshalling yards ; here they were engaged on purely local traffic. But the superheating of some of this class enabled them to work heavy coal trains between Colwick and the New England yard, Peterborough, a run of 47 miles. The saturated 0-8-2s were unable to undertake this turn as the tanks carried insufficient water to supply their increased consumption over that of the superheated engines. The fitting of superheaters to low-speed short-haul engines was not, however, inaugurated on these 0-8-2 tanks. In 1913 a new 0-6-0 shunting tank locomotive of Gresley's own design was put into service ; the first of this class, No. 157, used saturated steam, but a later one, No. 167 (put into service in 1914), was superheated. The respective leading dimensions are : cylinders, $18\frac{1}{2}$ in. $\times$ 26 in. ; wheels, 4 ft. 8 in. dia. ; working pressure, 175 and 170 lb. per sq. in. ; total heating surface, 980 and 932.5 sq. ft., the latter including 171 sq. ft. provided by the superheater ; grate area, 17.8 sq. ft. ; weight in working order, including 1,500 gal. of water and 3 tons of coal, $56\frac{1}{2}$ tons. These original engines are now L.N.E.R. Class " J51." They were the forerunners of the numerous standard 0-6-0 shunting tanks of the L.N.E.R., which are classed " J50," and are generally similar to the engine of 1913, except that the boiler is larger, 4 ft. 5 in. dia., against 4 ft. 2 in. in the original engine, with a total heating surface of 1,119 sq. ft. ; the grate area, however, is slightly less at $16\frac{1}{4}$ sq. ft. With a larger coal bunker to hold $4\frac{3}{4}$ tons, the " J50 " is slightly heavier, weighing 58 tons. It is not superheated.

In 1913 Gresley designed his twin-tube superheater. There are 34 flue tubes, 4 in. external dia., each element being in two flue tubes. By this arrangement greater evaporation and superheating areas are obtained ; the saturated header is located above the level of the flue tubes, and the superheated header below. How successful this layout proved in practice was shown in one of Gresley's last designs for the G.N.R., one that stands out from the general line of continuity in practice displayed in the locomotives built by him up to that time, and subsequently. The superheated 0-6-2 passenger tanks, the first of which was turned out in January, 1921, were just a simple straightforward

29

Gresley's twin-tube superheater

2-cylinder job, well suited to the pressing needs of the London suburban traffic. What appears to be the exceptionally high-line pitch of their boilers is rather an illusion caused by the severe restrictions of the Metropolitan loading gauge which cut the height of the engines working over the widened lines to 12 ft. 7 in., and made it necessary to use unusually squat boiler mountings. The leading dimensions of this class are as follows : cylinders 19 in. dia. by 26 in. stroke ; coupled wheels 5 ft. 8 in. dia. ; total heating surface, 1,205 sq. ft., of which the 17-element superheater

30

Photo]   *[W. J. Reynolds*

Ivatt 0-6-2 tank No. 1598 of 1912, fitted with superheater by Gresley, and tried out before the building of the Gresley " N2 " tanks. Now L.N.E.R. No. 4598 Class " N1 "

contributes 207 sq. ft. ; grate area, 19 sq. ft. ; working pressure, 170 lb. per sq. in. ; weight in working order, 70¼ tons, including 2,000 gallons of water and 4 tons of coal. Such preliminary

Photo]   *[W. J. Reynolds*

Gresley's superheater 0-6-2 tank for London suburban services ; No. 4750, L.N.E.R. Class " N2 "

31

Ivatt " N1 " 0-6-2 tank armoured for coast defence train in 1914

Ivatt 0-6-0 equipped by Gresley for overseas service in the 1914-19 war

32

24

experience as was necessary had already been obtained by the superheating of one of the numerous Ivatt 0-6-2 tanks (L.N.E.R. Class " N1 ") No. 1598, and the new locomotives, which were a logical development of Ivatt's design, were drafted to the most arduous suburban duties the moment they were broken in.

Sixty of these engines, now L.N.E.R. class " N2," were turned out very rapidly, 10 from Doncaster, and 50 by the North British Locomotive Co. Ltd., the original numbers being 1606-15 (Doncaster) and 1721-70. Further examples, without condensing apparatus, have been built since grouping for suburban service around Glasgow and Edinburgh. They are not only powerful engines for their size, but also speedy. Before the days of the Pacifics they were sometimes requisitioned to pilot main line expresses out to Potters Bar ; while their hill-climbing feats, over the tremendous gradients of the High Barnet branch, though a commonplace today, were exceptional when they first came out. Up the 2½ miles at 1 in 59-63 from Finsbury Park, for example, with the trains that pass a number of stations, speeds usually rise to 30 m.p.h., or so, with loads of 160 tons.

Mention of the Ivatt 0-6-2 tanks in connection with the experimental work prior to the building of the " N2s," recalls the equipment by Gresley in 1915 of two of them with armouring for use in coastal armoured trains. They could be driven from the footplate or from either end of the train in which they were placed in the middle. Further special locomotive work, carried out at Doncaster during the war of 1914-18 under Gresley's direction, involved the modification of a number of 0-6-0 tender engines (L.N.E.R. " J3 " class) so that they could exhaust through the chimney, behind the tender, or condense into the tender. Thus equipped, they were used in forward areas in France.

The first of a new batch of 3-cylinder 2-8-0s came out in 1921 ; they differed from No. 461 in having the alteration to the derived valve-gear that had proved so satisfactory in the " 1000 ". class Moguls. No. 477 was the first of the new series, now L.N.E.R. Class " O2," in which the principal dimensional changes from No. 461 were an increase in boiler pressure from 170 to 180 lb. per sq. in., and the enlargement of the cylinders from 18 to 18½ in. dia. The cylinder disposition is almost identical with that of the " 1000 " class, the inside cylinder being steeply inclined so that the three steam chests can be arranged in line. This type is now the standard heavy freight engine of the L.N.E.R.

33

Photo]   *[W. J. Reynolds*

No. 3479 (*originally* G.N.R. 479), *one of the first 3-cylinder* 2-8-0s, *Class* " O2 "

(Note the difference at the front end from No. 461, see page 23)

While these new types were being introduced, an important modification to the large Atlantics was being applied one by one to the whole class. The ten superheater engines built by Ivatt, Nos. 1452-1461, showed a definite though not very great superiority over the saturated variety, particularly in uphill work, but their work fell considerably short of the feats to which we are now accustomed. They were the victims of a mistaken conception of superheating. Ivatt attempted to exploit the principle purely for the reduction of boiler maintenance, reducing the working pressure from the 175 lb. of the saturated engines to 150 lb. per sq. in., so that although the cylinders were increased in diameter from 18¾ in. to 20 in. the power of the engines was barely altered. There were steaming troubles, too—a thing almost unheard of with the " 251s " of today, or with the original non-superheater engines. When a start was made with the superheating of the original Ivatt Atlantics bolder measures were taken, the boiler pressure being retained practically at the former figure of 175 lb. per sq. in. The process of transformation was gradual, for many of the engines at first retained their slide-valves, while others were fitted with new 8-in. dia. piston-valves. The first engines to be equipped were provided with 24-element superheaters, having the same amount of heating surface—427 sq. ft.—as Nos. 1452-61 and the 4-cylinder engine No. 279. In 1919, however, No. 1403 was fitted with a 32-element superheater, having 568 sq. ft. of heating surface. This variety later became standard for the whole class, and those engines which for a time had

34

24-element superheaters were subsequently modified. By no
means all of them have been fitted with piston valves, however ;
the slide-valve engines, which retain their 18¾ in. cylinders, are
now mostly to be found on the Great Central section.

Credit for the wonderful work of the " 251 " class is usually
bestowed in its entirety upon Ivatt, but the modifications that
transformed them from a type of moderate and at times indifferent
performance into some of the most capable engines of their size
and weight that have ever run on British metals were carried
out during the Gresley *regime.* The piston-valve engines, with
20-in. cylinders, have done the finest work, but before the present
war the Sheffield drivers used to get excellent running out of
their slide-valve engines. It was during the later years of the
1914-1919 war that the full haulage capacity of the superheated
" 251s " was first demonstrated, and although those early feats,
by the 24-element superheater engines, have since been totally
eclipsed by the still more capable " 251s " of today, on very much
faster schedules, one can quite well imagine the thrill of a run
like that of No. 1407, which, with the 1.40 p.m. down and a load
of 575 tons, ran from Potters Bar to Peterborough, start to stop,
63.7 miles in 74 minutes. Still finer was the 1921 coal strike
exploit of No. 290 in working a load of 600 tons from Peterborough
to Kings Cross, 76.4 miles, in 92 min. 35 sec. What is more, they
were never piloted, save from Kings Cross to Potters Bar ; this
luxury, however, was permitted only when the load exceeded a
modest 66 axles.

Photo]   [L.N.E.R.
No. 3833, one of the latest Class " O2 ", 3-cylinder type, built by
L.N.E.R., with enlarged cab and standard tender

35